Network Logistics Games

Logistics Games

Design and Implementation

SYDNEY LITTERER, JENNIFER BROOKES, STEPHEN M. WORMAN,
DAVID A. SHLAPAK

Prepared for the Joint Staff J8
Approved for public release; distribution unlimited

NATIONAL DEFENSE RESEARCH INSTITUTE

For more information on this publication, visit **www.rand.org/t/RRA470-2**.

About RAND

The RAND Corporation is a research organization that develops solutions to public policy challenges to help make communities throughout the world safer and more secure, healthier and more prosperous. RAND is nonprofit, nonpartisan, and committed to the public interest. To learn more about RAND, visit www.rand.org.

Research Integrity

Our mission to help improve policy and decisionmaking through research and analysis is enabled through our core values of quality and objectivity and our unwavering commitment to the highest level of integrity and ethical behavior. To help ensure our research and analysis are rigorous, objective, and nonpartisan, we subject our research publications to a robust and exacting quality-assurance process; avoid both the appearance and reality of financial and other conflicts of interest through staff training, project screening, and a policy of mandatory disclosure; and pursue transparency in our research engagements through our commitment to the open publication of our research findings and recommendations, disclosure of the source of funding of published research, and policies to ensure intellectual independence. For more information, visit www.rand.org/about/principles.

RAND's publications do not necessarily reflect the opinions of its research clients and sponsors.

Published by the RAND Corporation, Santa Monica, Calif.
© 2023 RAND Corporation
RAND® is a registered trademark.

Library of Congress Cataloging-in-Publication Data is available for this publication.

ISBN: 978-1-9774-1-0870

Cover photo by U.S. Army.

About This Report

In wargames, the logistics necessary to support fielded forces are frequently abstracted or ignored altogether because of the complexity of logistics systems and the need to ensure the manageability of game rule sets. Logistics are instead typically studied as an optimization problem. However, ignoring logistics is paradoxical given the importance of resupplying units while waging and winning wars in the real world, and studying them through optimization omits the complicated dynamics of human decisionmaking in crises—something that wargames are well suited to explore. To address the dearth of logistics-focused wargames, this report details a logistics game design that reflects the complexity of logistics systems without requiring computer aids. The report is intended to guide both experienced and inexperienced wargame designers in creating a game of this type to answer their own research questions.

This work was funded by the Joint Staff J8 (Force Structure, Resources, and Assessment Directorate) as part of a broader portfolio of wargaming work that began in 2020. As part of that work, RAND researchers executed a game based on this design in fall 2021. However, this report does not describe the specifics of that game. Rather, it provides a detailed description of the principles followed in creating the game design, the typical game components, the steps for adapting this game design to answer a given research question, and the process of executing a game using this design.

The research reported here was completed in November 2022 and underwent security review with the sponsor and the Defense Office of Prepublication and Security Review before public release.

RAND National Security Research Division

This research was sponsored by the Joint Staff J8 and conducted within the Acquisition and Technology Policy Program of the RAND National Security Research Division (NSRD), which operates the National Defense Research Institute (NDRI), a federally funded research and development center sponsored by the Office of the Secretary of Defense, the Joint Staff, the Unified

Combatant Commands, the Navy, the Marine Corps, the defense agencies, and the defense intelligence enterprise.

For more information on the RAND Acquisition and Technology Policy Program, see www.rand.org/nsrd/atp or contact the director (contact information is provided on the webpage).

Acknowledgments

We would like to thank our colleagues, Kelly Eusebi, David Frelinger, Bradley Martin, Bryan Rooney, and Emily Yoder, for their support throughout the process of designing and implementing the original game on which this report is based. We are also grateful to the Joint Staff J4 (Logistics Directorate) for its contributions to that process, as well to the Joint Staff J8, which sponsored this work. Finally, we appreciate the assistance that our reviewers, Sarah Smedley and Ellie Bartels, provided in improving this document.

Summary

In wargames, the logistics necessary to support fielded forces are frequently abstracted or ignored altogether because of their complexity and the need to ensure the manageability of game rulesets. Abstracting logistics is paradoxical given the importance of resupplying units when waging and winning wars in the real world, and studying them through optimization omits the complicated dynamics of human decisionmaking in crises—something that wargames are well suited to explore. This report details a logistics game design that reflects the complexity of logistics systems without requiring computer aids, which can be too time-consuming to use during game adjudication and may require technology that is not available in secure game venues. The design is appropriate for gaming contested logistics involving an active adversary, as well as logistics under other challenging conditions, such as evacuations and movement of aid after a natural disaster. One of the key benefits of using a game for this purpose is that games enable exploration of the interaction between friendly and adversary decisionmaking and provide insight into how players respond to operational surprise.

The game design challenges players to meet demand for a given set of supplies by deploying transportation assets across a geographically dispersed network of locations, which we describe as hubs (supply distribution locations) and spokes (locations with supply demand).[1] During gameplay, players must balance the resources available to transport supplies with the demand generated by the spokes, make choices about how to allocate different transportation assets, determine which locations to prioritize in the event of shortfalls, and develop strategies to mitigate disruptions. This game design can be used for a variety of purposes, including the following:

- implementing various concepts for logistics operations and exploring the potential feasibility and possible implementation challenges of those concepts

---

[1]   In the context of hub-and-spoke models, the term *spoke* can refer to either a route leading out from a hub location or to the endpoint of that route. This report uses the term in the latter sense.

- generating insight into the vulnerability of a logistics network to disruption from either natural or adversarial causes
- understanding logistics system dynamics in the context of unpredictable human decisionmaking
- generating solutions to problems encountered in real logistics systems
- improving combat-focused wargames by identifying conditions under which it is or is not reasonable to assume forces will have adequate supplies to carry out operations.

Gameplay and Adjudication

At the beginning of the game, the players are given a scenario that requires the Blue team to transport supplies to various locations shown on a map in order to meet as much demand as they can. The Blue team members are told what type of resources they can use as they attempt to meet their objectives. If a Red team is playing, these players are also briefed on their objectives, which typically involve preventing or altering Blue supply movements. At this point, the game begins, and turns proceed according to the following process:

1. The Blue and Red teams separate to plan their first move before communicating their intent to the adjudication team.
2. The adjudication team uses this intent to determine how assets and supplies move throughout the transportation network and resolve conflicts between Blue and Red actions. The adjudicators then calculate how well demand has been met at various locations and record relevant data. The initial movement of supplies throughout the network is adjudicated according to various game rules; adjudication of conflicts between Red and Blue actions relies on adjudicators' subject-matter expertise.
3. The adjudication team briefs the players on Blue and Red team actions during the turn, amount of demand met at each location, and any other notable changes.
4. The game board is reset as needed, and players start planning their next turn.

When the game is completed, a hotwash session is typically conducted to discuss lessons learned during the game. Data collection is conducted throughout the game; the most important data will typically consist of player discussions and decisionmaking, the levels of demand met by players at each location on each turn, and what resources are available to players during each turn.

Game Components

The game is played using two maps; playing pieces representing transportation assets, personnel, and supplies; and various chips indicating such information as locations of obstacles or Red attacks. The first map used is a geographic map depicting logistics network hubs and spokes with associated demand per move. The second is a detailed network diagram intended primarily to allow the adjudication team to track the movement of assets and supplies throughout the transportation network and calculate met demand. These maps are shown in Figure S.1.

Transportation assets, operators, and supply game pieces each represent some number or volume of transportation assets, personnel, or supplies. Additional required game pieces will depend on the game scenario and will

FIGURE S.1

Notional Geographic and Network Maps Showing Locations and Demand per Move

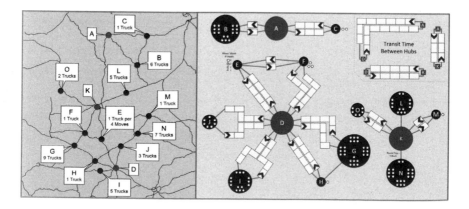

typically indicate or track information about the logistics system that could change as the game is played.

Game Customization and Preparatory Analyses

This game can be tailored to new scenarios in five broad steps: developing the game scenario, determining how to represent demand across a network of locations, calculating the number of transportation resources that will be available to players, creating customized game rules, and conducting supplementary preparatory analyses. These components should be thought of as interlocking processes rather than sequential stages of design, and completing them will likely require some iteration.

Scenario development provides the scope and detail necessary to develop other game components, so it should be considered from the beginning of game design. Care should be taken to create scenarios that are relevant and provide enough context to help players frame their decision space. This game is most appropriate for scenarios in which the Blue team will face unexpected obstacles, transportation resources are not abundant relative to demand, and the logistics network is relatively fixed. These conditions force players to make the types of decisions that games are helpful in exploring and ensure that the outcomes of those decisions can be adjudicated using the game rules. If the Blue team will not face challenges, such as unexpected adversary interference, environment obstacles, or resource shortfalls, then the game will not be particularly useful. If the transportation network (the locations of hubs and spokes and the most commonly used routes between them) is highly dynamic, this game design will be difficult to use because adjudication relies on a physical representation of the network that cannot be easily updated during gameplay.

After the scenario is developed, the geographic and network maps are created by specifying which hubs, spokes, and routes are relevant to the game, determining how long it will take for transportation platforms to move along each route, and determining how much demand exists at each spoke location per turn. The process for determining players' supply of transportation assets and operators requires determining how many assets or operators a given physical game piece represents, identifying relevant categories of assets and operators, and using the previously determined information to calculate how many game pieces players should receive for each category. One of the overarching concerns when tailoring this game to a new scenario is the need to represent reality well enough to meet the game objectives without making the information players need to process and the decisions they need to make too complex.

Game rules govern how player decisions are implemented and adjudicated by the adjudication team and should rely on research into how the various game components function. In this game, the rules primarily specify how transportation platforms, operators, and supplies are allowed to move around the network map during the adjudication process. Additional analyses may be required to adjudicate the impact of obstacles or disruptions on the system, the success or failure of players' strategies for preventing or overcoming obstacles or disruptions, and creative player choices.

Conclusion

This game design is flexible enough to accommodate a variety of scenarios, objectives, and modifications. It lends itself particularly well to scenarios in which resources are limited or unexpected obstacles could appear, including nonmilitary scenarios. Its potential uses include exploring new concepts of operations, illuminating vulnerabilities in logistics networks, understanding the dynamics of supply movements, highlighting the broader impacts of logistics planning, and brainstorming solutions to challenges. The game design could be modified to accommodate networks that are highly interconnected, to incorporate detailed mechanisms governing Red team actions, to consider both Red and Blue logistics systems, and to integrate this game into a series of other wargames. However, future iterations of this game should preserve the simple, transparent, and tangible way in which logistics networks are represented within the game by balancing the fidelity with which logistics are represented with ease of play, enabling players to engage with the problems presented by the game at whatever level of detail they are most comfortable with, and allowing players to explore innovative solutions to the problems they have been given. Maintaining these key strengths of the game design will allow it to remain a useful and impactful tool, even as it is adapted to answer new questions and explore new possibilities.

Contents

Figures and Tables

Figures

Tables

CHAPTER 1

Introduction

It is seemingly paradoxical that (1) historically, logistics are often ignored or abstracted at a very high level during operational wargames and (2) the ability to resupply units is vital for waging and winning a war in the real world. While attention to logistics has been increasing in recent years, there is still much to be done in improving how logistics and logistical concerns are represented via wargames.[1] Traditionally, the study of logistics fundamentally seeks to solve the problems associated with the movement of specified commodities from Point A to Point B and is typically treated as an optimization problem. When assumptions about the exact unit operating in a given location and what precise missions will be executed on what time frame are unavailable, the ability to do detailed adjudication of logistics shortfalls in real or near-real time is very limited. Furthermore, logistics is often of ancillary importance to the objectives of the wargame; that being said, there have been many games in which logistics are the primary focus, and others in which designers have sought to increase the fidelity with which logistics are represented.[2]

This report details a methodology in the tradition of increasing fidelity, developed to explore logistics either as a stand-alone game or played as a companion to a more operationally focused game. This means that the game focuses on the movement of supplies to meet demand rather than the impact of logistics on broader operations, but integrating the outcomes of

---

[1] For some open-source examples of how logistics has been the focus of, or incorporated into, previous wargames, see Donnelly et al., 2016; Krievs, 2015; LaPlante, Garner, and Hutzler, 1995–1996; and Mays et al., 2017.

[2] For an example of how assumptions regarding logistics might have affected the results of at least one recent wargame, see Vershinin, 2021.

this game into broader operational wargames is one potential avenue for future work. Part of the impetus for the creation of this game was the need to balance the complexity of logistics in a way that is manageable without computer aids.[3] Often, logistics subject-matter experts (SMEs) provide inputs into the adjudication process to help identify potential shortfalls or ways in which concepts of operation (CONOPSs) might fail in the real world if steps to alleviate these potential shortfalls are not undertaken. One example of this is considering aerial refueling requirements for fighter aircraft and bomber treadmills. By forcing aviators to consider the limitations on operations, higher-quality results are theoretically obtained. The downside to this approach of asking for generalities on aerial refueling needs and the potential impact on operations is a high degree of imprecision and difficulty in quantifying the actual operational impacts. The game design detailed in this report seeks to address this class of shortcoming.

The game design challenges players to deploy transportation assets with the goal of meeting demand for a given set of supplies across a geographically dispersed network of distribution points, which we describe as hubs and spokes. Hubs are locations from which supplies are distributed; spokes are locations with demand for those supplies.[4] During gameplay, players must attempt to balance the resources available to transport supplies with the demand generated by the spokes, make choices about which locations to prioritize in the event of shortfalls, and develop potential strategies to mitigate adversary interference or other disruptions. For instance, players could lay out detailed plans to vet contractor crews assigned to high-priority missions. Alternatively, they may suggest positioning security forces along critical routes. This game design can be used for a variety of purposes, including the following:

---

[3]   Classification issues often limit access to computers throughout game play, and running computer models as part of game adjudication can be impractical given typical time constraints.

[4]   In the context of hub-and-spoke models, the term *spoke* can refer to either a route leading out from a hub location or to the endpoint of that route. This report uses the term in the latter sense.

- implementing various concepts for logistics operations and exploring the potential feasibility and possible implementation challenges of those concepts
- generating insight into the vulnerability of a logistics network to disruption from either natural or adversarial causes
- understanding logistics system dynamics in the context of unpredictable human decisionmaking
- generating solutions to problems encountered in real logistics systems
- improving combat-focused wargames by identifying conditions under which it is or is not reasonable to assume that forces will have adequate supplies to carry out operations.

This game is most appropriate for scenarios in which the Blue team will face unexpected obstacles, transportation resources are not abundant relative to demand, and the logistics network is relatively fixed. These conditions force players to make the types of decisions that games are helpful in exploring and ensure that the outcomes of those decisions can be adjudicated using the game rules. If the Blue team will not face challenges, such as unexpected adversary interference, environment obstacles, or resource shortfalls, then the game will not be particularly useful. When all challenges are known ahead of time, meeting demand is simply a problem of optimization that could be handled by modeling and simulation; when Blue has enough resources to meet demand regardless of any likely obstacles they could face, the stakes of the game are very low. If the transportation network (the locations of hubs and spokes and the most commonly used routes between them) is highly dynamic, this game design will be difficult to use because adjudication relies on a physical representation of the network that cannot be easily updated during gameplay.

A version of this game was created and executed for the Department of Defense Joint Staff, J4 (Logistics Directorate), in September 2021. This report details the general game design and provides insight into how other games might incorporate similar designs. As a stand-alone game, the methodology detailed in this report is most appropriate for exploring operational- or tactical-level problems that involve moving a given class of supplies through a relatively static network of hub and spoke locations to meet a fairly predictable level of demand. RAND's September 2021 game looked at contested logistics and included an active Red team seeking to interfere with Blue supply movements, but this design is also appropriate for incorporating disruption from sources such as weather, traffic, or supply shortages.

The rest of this chapter is meant to provide context for the rest of the report. We first describe how a hypothetical game might be run. This provides an example of the type of game design that can be built using the process described in Chapter 3, as well as a preview of the game play and adjudication discussed in Chapter 4.[5] We then outline the key principles guiding our game design process. While these principles are common to a variety of games, we describe them in detail in this report for the benefit of readers new to game design.

Game Day Preview: Logistics in Other Countries

This section describes a hypothetical game using the design outlined in this report.

A group of players affiliated with the Department of Defense meet at a RAND facility for a one-day game exploring potential vulnerabilities in conducting ground-based logistics in allied countries (Country X, Country Y, and Country Z) during a war between two near peers. The players bring a mix of expertise on military logistics, the theater, and adversary doctrine. At the beginning of the game, the players are introduced to the scenario. In particular, they are provided an overview of adversary and allied military activity and told that allied operations have created region-wide demand for a certain consumable commodity typically transported by truck. At this point, the players' attention is directed to a large map on a table in the center of the room showing supply distribution hubs and the locations with demand for supplies (see Figure 1.1). Supply and demand for each game turn are indicated on tokens placed at each location and given in terms of the number of truckloads of supplies required.

---

[5] The scenario described here is notional, and this game design can be adapted to a wide variety of scenarios that might look very different from the one described in this chapter.

FIGURE 1.1

Notional Geographic Map Showing Locations and Demand per Move

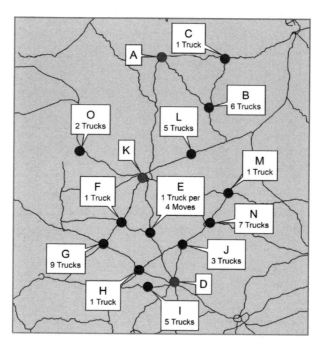

After the game map is introduced, the adjudication team describes the resources available to the Blue team for transporting assets, providing team members with a handout listing the number and types of transportation platforms and platform operators available across the region at the start of the first turn of the game. In the game scenario, only military-owned trucks and military personnel are available to move resources. Table 1.1 shows the information provided on the handout.

Players assigned to the Blue team are then told their responsibility for each game turn will be to determine how they want to prioritize meeting demand at the locations shown on the game map. The players are also asked to decide what measures they will take to prevent adversary interference with allied supply movements, which could include measures protecting transportation platforms, platform operators, supplies, storage facilities, and transportation infrastructure. Because the teams represent command-

TABLE 1.1

Information on Player Handout

Nationality	# of Trucks Available	# of Truck Drivers Available
Country X	32	24
Country Y	24	16
Country Z	12	20

NOTE: Notional data.

ers in charge of logistics operations, these are treated as "requests for protection" or "requests for defensive assets" by the players to the combatant commander. Players will need to make a compelling argument as to why an asset should be pulled off one mission to prioritize theirs. This prevents players from over- or unrealistically allocating assets and also allows the adjudication team to potentially track what was requested for future study. Players assigned to the Red team are similarly told to determine what measures they will take to interfere with Blue supply movements during each turn.

After each team understands their responsibilities, players are given a brief overview of how their decisions will be adjudicated at the end of each turn. The diagram shown in Figure 1.2 is displayed; this diagram represents an abstract version of the game map the players will be using, with additional details regarding demand requirements, typical transportation routes, and transit times. The adjudication team will use physical game pieces representing transportation platforms, operators, and supplies to track the movement of supplies throughout the transportation network based on player priorities for meeting demand and to calculate how much demand has been met at each location at the end of each turn.

Once the game is introduced, the Red and Blue teams move to different rooms and discuss their first turn. Once each team is finished deliberating, the teams reconvene and describe their proposed priorities and actions to the group. The Blue team states that they intend to prioritize fully meeting demand in Country X while ensuring that at least half of the stated demand is met in locations in Country Z and Country Y. They intend to safeguard trucks during transit by providing each with an armed escort. They also decide to provide additional security at storage facilities at the locations sup-

FIGURE 1.2

Notional Network Diagram Used for Adjudication

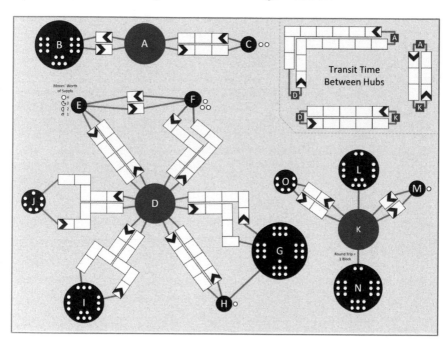

plies are traveling to. The Red team states that they want to destroy supply storage facilities using special operations forces and damage key bridges along major transportation routes using missile strikes.

The adjudication team then convenes around the adjudication diagram shown in Figure 1.2 and—guided by the Blue team's priorities—moves physical game pieces representing platforms, operators, and supplies along the routes shown on the diagram according to predetermined game rules. Once this is complete, the adjudicators modify the progression of pieces across the board to reflect the results of Red actions, as determined by adversary SMEs on the adjudication team. This process includes moving pieces back multiple spaces to reflect delays and removing supply pieces that were unloaded at destination points to reflect destroyed supplies.

Next, the adjudication team counts the units of supplies that have been successfully delivered to each location and reports back to the Red and Blue teams on the results of their actions, noting that, while the Blue team successfully met demand in Country X, they were able to meet only 25 percent of demand in Country Y and Country Z, potentially affecting the ability of allied forces to operate from within these countries. This shortfall occurred in part because the Blue team did not have enough transportation platforms available and in part because platforms traveling certain routes in Country Z had to take time-intensive detours after adversary missiles damaged bridges. The players are also told that storage facilities in Country Y were targeted by adversary saboteurs, but, because of an increased security presence, the adversary succeeded in destroying only a portion of supplies at one facility.

With the results of the previous turn in mind, the Red and Blue teams once again break out into different rooms to plan for Turn 2. The Blue team is unsatisfied with the amount of demand met in Country Y, so they redirect some platforms from destinations in Country X to destinations in Country Y at the cost of failing to completely meet demand in Country X in the next turn. However, they are worried that the Red team will shift the focus of future missile attacks to Country X, so they devote missile defense assets to protecting Country X's infrastructure to avoid losing even more capacity to meet demand in Country X. Because saboteurs targeted storage facilities at spokes in Country Y, Blue continues their elevated security measures at spoke storage facilities and increases security at hub storage facilities. The Red team is pleased with the results of their attacks in Country Z infrastructure but realizes that the Blue team's logistics efforts are centered in Country X, so they decide to direct their next missile attacks to critical transportation infrastructure in Country X. Because their attacks on storage facilities were less successful than they had hoped, the Red team decides to cease those attacks and instead attempt to bribe Blue truck drivers to sabotage their own vehicles.

The cycle of planning and adjudication is repeated multiple times before the end of the day, when the players and adjudication team meet for a hotwash session. During the hotwash, the players discuss what they learned during the day, focusing—per the game's objectives—on Blue vulnerabilities that players found particularly unexpected or difficult to overcome.

Design Principles

In this section, we describe three key principles that drove the development of the network logistics game: transparent adjudication, layered complexity, and flexible rules. *Transparent adjudication* means ensuring that players can understand how the decisions they make produce certain results, as well as how the game mechanisms both reflect and abstract from reality. Doing this builds players' trust in the game and can reduce frustration about negative outcomes. However, requiring players with significant policymaking and subject-matter expertise but little wargaming experience to actively apply detailed rules is often a poor use of their time, which underscores the need for layered complexity. For that reason, it is useful to allow players to make moves according to less complex rules than those used to adjudicate the results of those moves. When players are making their moves, it is not uncommon for them to propose actions that are not explicitly covered by the game rules, but that are nonetheless realistic given the game's scenario. Ensuring that both the game's rules and adjudication are flexible enough to accommodate this creative decisionmaking further bolsters the game's credibility. It is important to note that these design principles build on many best practices in the field.[6]

Transparent Adjudication

When designing the logistics game, we prioritized developing adjudication procedures that were transparent and easy to understand for both the adjudication team and the players. Transparent adjudication procedures improve player buy-in; even if players are unsatisfied with the outcome of adjudication, they can understand how their inputs led, directly or indirectly, to the resulting outcomes (Weuve et al., 2004). Transparent adjudication also increases the potential scope of a game's learning objectives. When players can see a clear connection between their decisionmaking and the consequences of their decisions, there are opportunities for them to identify

---

[6] As noted in the text, these design principles are generally considered best practices in the field, and brief descriptions of each are included here for the ease and convenience of the reader. For more thorough examinations of many of these topics, see Bartels, 2020; Caffrey, 2019; Hanley, 1991; Perla and McGrady, 2011; Perla, 1990; and Rubel, 2006.

and discuss solutions for the root causes of adverse adjudication results. For this game, transparent adjudication is achieved via tracking the movement of resources on a physical diagram using simple, easy-to-follow rules representing real-world constraints. Players are free to engage with this diagram and scrutinize the adjudication process throughout the game.

Layered Complexity

Serious games conducted by RAND are often intended for players who have very little time to participate in wargames. Therefore, we want players to spend most of their time contributing their expertise rather than learning detailed game rules. At the same time, logistics systems are often complicated, and it is important to maintain an appropriate level of fidelity when representing them to ensure that players are faced with a realistic problem (Bartels, 2020). Our network logistics game balances these two considerations by allowing players to operate at a different level of complexity than the adjudication team. During the game, players use high-level information about the logistics network to make decisions about their priorities and general strategies, which they communicate to the adjudication team. The adjudication team translates player guidance and expressed intent into detailed movement of resources through the transportation network to calculate how much demand is met in a given turn. These two levels of complexity manifest themselves in the materials used to play the game—the players use a map of the area of interest with relevant hub and spoke locations and their associated demand indicated, and the adjudication team uses a detailed diagram on which game pieces are used to represent individual movements of defined quantities of supply between specific points. Players may wish to engage with the more detailed hub-and-spoke diagram when working through particularly tricky problems and should be allowed to do so if it supports the game's objectives.

Flexible Rules

While designing game rules and materials for this game requires making certain simplifying assumptions about the logistics system, it is important to make sure that players can still make realistic decisions during the game. In other words, the design of rules and materials should not limit players' decision space within the scope of the game (Bartels, 2020; Weuve et. al., 2004). For example, the adjudication diagram used in the previous iteration of this game conducted by RAND depicted a network of hubs and spokes with typical routes between hubs and spokes explicitly shown on the game board, but it would have been reasonable for players to redirect transportation assets from a certain hub to a spoke location not typically connected to that hub. Therefore, we conducted an analysis of the transit time between every hub-and-spoke combination to ensure that we could accommodate that type of decision, though we did not necessarily provide this information to the participants. We did not, however, determine transit time for locations outside the logistics network, because travel to those locations was not within the scope of the game. In general, we made sure that game rules, materials, and adjudication processes were flexible enough to adapt to creative player choices. The purpose of these flexible game rules was to avoid artificially constraining the players' decision space while minimizing complexity in game play and adjudication.

Integrating These Design Principles

Aligning the game design with the principles described above allows the game to remain player-friendly while credibly depicting complex military logistics. The principles of transparent adjudication and layered complexity go hand-in-hand. In this two-level game design, the adjudication team converts players' high-level guidance into detailed supply movements, then reports back progress. Without any visibility into how this adjudication is done, players can feel as if their decisions are disconnected from the eventual results, particularly if they expected a different outcome than the one they got. Providing the players with visibility into the adjudication process mitigates this risk without sacrificing the benefits that a layered design provides.

While there is very little tension between the principles of transparent adjudication and layered complexity, they can be challenging to balance with the need for flexible rules, particularly in this game design. The use of game pieces on a physical diagram to track supply movements enables both layered complexity and transparent adjudication; however, a physical diagram can be difficult to modify in response to unexpected player decisions, and players' familiarity with the logistics system model could influence the decisions they make, particularly if the players are inclined to "game the system" to get the results they want. The degree to which this tension will affect a game using this design will depend on its scope and objectives; balancing all three design principles requires careful consideration of the decisions players are expected to make and how particular player behaviors will work toward or against the game's goals.

Report Overview

The rest of this report provides additional details on the design and implementation of a network logistics game before concluding with a discussion of how this game design could be used in the future. Chapter 2 provides an overview of the various physical components of the game. Chapter 3 describes the process of designing those game components, as well as the rules of the game. Chapter 4 lays out how to conduct gameplay and adjudication. Chapter 5 concludes the report and describes potential uses for this game design, as well as ways in which it could be modified. The information in these chapters is relatively detailed because it is intended to be helpful to both experienced and inexperienced game designers.

CHAPTER 2

Game Components

This chapter details the individual components required for the logistics game. This particular game is played using two maps; playing pieces representing transportation assets, personnel, and supplies; and various chips indicating such information as locations of obstacles or Red attacks. Each component is described briefly in this chapter. The specific materials required may vary with application, and we will note some examples below.

Maps

In keeping with the principle of layered complexity, this game is played on two maps. The first map, which we will refer to as the geographic map, is a high-level depiction of locations and their associated demand set against real geography (see Figure 2.1 for a notional example). The geographic map is meant to provide the basic information players need to determine how to prioritize the distribution of assets—specifically:

- geographic arrangement of hub and spoke locations
- demand per turn at spoke locations
- distance between hub and spoke locations.

Providing a geographic map makes the scenario more concrete for the players because they are able to see the theater in the way they are most familiar with, and it ensures that only the most relevant information is displayed to avoid overloading them with information. This aligns with the principle of layered complexity discussed in Chapter 1. In the original game design, knowledge of the "generally anticipated" course of the conflict and

FIGURE 2.1

Notional Geographic Map Showing Locations and Demand per Move

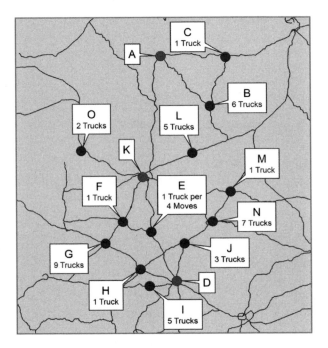

its attendant logistical challenges was a selection criterion of the players and discussed with players if and when questions arose. While no explicit disagreements on assumptions about CONOPSs arose, if they do, the first question to ask the players would be whether that difference would manifest as a significant difference in the logistics operations on the board. If so, the adjudication team can consider adding or removing supply spaces on the board if necessary.

The second map is a detailed network diagram (see Figure 2.2 for a notional example). The network diagram's purpose is to track the movement of assets and supplies throughout the transportation network and to allow the adjudication team to determine how well players are meeting demand. It essentially acts as a physical calculator, simulating the dynamics of the transportation system in a detailed yet transparent manner. This type of computation could also be completed using a computer simulation;

FIGURE 2.2

Notional Network Diagram Used for Adjudication

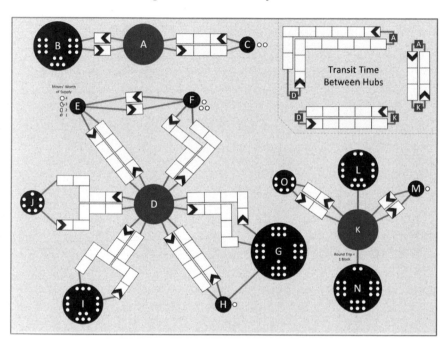

however, the physical method is often preferable because it allows players to easily grasp the inner workings of the model. Enabling this understanding can be invaluable in stimulating problem-solving and discussion, particularly for research games. This map was originally designed primarily for utilization by the adjudication team, given concerns over the complexity of its design. There is no reason this map cannot be given to players, but, in our estimation, it is best to introduce players to the geographic map first and introduce the more detailed map only if necessary. This prevents players from being overwhelmed with details, especially if the goal of the game is to focus on strategic decisions. Providing the network diagram to the players can be useful if, for example, players are struggling to consistently meet demand even though the total capacity of available transportation platforms exceeds the total demand required each turn. In this situation, examining the movement of game pieces across the network diagram could help them see that they need to assign multiple transportation platforms to longer routes so that a new platform arrives every turn.

The network diagram provides the following information:

- abstracted arrangement of supply hubs and destinations (hubs and spokes)
- time required to complete each route
- demand per turn at each spoke (may be more detailed than demand shown on high-level map).

Route completion time and demand for supplies are each represented in discrete increments. Choosing appropriate increments for each is covered in Chapter 3.

Transportation Platforms, Operators, and Supplies

Three types of playing pieces are used in conjunction with the network diagram to perform adjudication. These are pieces which each represent some number or volume of transportation platforms, personnel, or supplies. For transportation platforms, one physical playing piece represents a specific number of a certain type of vehicle and has a specific supply carrying capacity: For example, one game piece represents two semitrucks, which together can carry 50 pallets. For personnel, one playing piece represents a specific number of transportation platform operators: For example, one game piece represents two truck drivers. For supplies, one game piece represents a certain volume or weight of a specific type of supply: For example, one game piece represents 25 pallets of Class V supplies.[1]

Operators and platforms are depicted separately primarily because operators introduce different vulnerabilities to the logistics system than platforms do. For example, in a contested logistics scenario, operators could be convinced to sabotage platforms or contaminate supplies. In a humanitarian scenario (e.g., natural disaster or pandemic), operators could become unable to perform their duties because of sickness or injury. The vulnerabilities associated with operators, platforms, and supplies can differ based on the characteristics of those game components (e.g., operators who

---

[1] For a description of U.S. supply classes, see Army Doctrine Publication 4-0, 2019.

receive less vetting could pose a higher security risk). Because of this, it can be useful to categorize platforms, operators, and/or supplies based on relevant characteristics—color-coding the playing pieces accordingly. A non-exhaustive list of potential categorizations is provided below:

- Transportation Platforms
  - transportation mode (e.g., land, air, sea)
  - supply capacity (e.g., 10 pallets, 25 pallets, 50 pallets)
  - speed (e.g., 15 knots, 25 knots)
  - owner (e.g., Department of Defense, Company A, Company B)
  - nationality (e.g., American, French, German)
- Personnel
  - nationality (e.g., Japanese, Australian, Indonesian)
  - qualifications (e.g., has or does not have Commercial Driver's License)
  - vetting status (e.g., has or has not completed a background check)
  - military status (e.g., military, civilian)
- Supplies
  - military supply class (e.g., Class III, Class V, Class VII)
  - handling requirements (e.g., hazardous material, nonhazardous material).

Determining how many categories of transportation platforms, personnel, or supplies to include in the game is largely a matter of balancing the realism of the game with its simplicity. In games, players are asked to make complex decisions. This process can be made easier if the rules and components of the game are simplified whenever possible to reduce the amount of information players need to remember and process. However, a game still needs to model reality with enough fidelity to answer its associated research questions. This limits how far the game can be simplified. As an example, in a game designed to understand how Red might exploit platform and operator vulnerabilities to disrupt logistics, it would be helpful to categorize operators and platforms by characteristics that background research suggests might be relevant, such as platform owner or operator vetting status. However, the research question can be adequately answered without including different types of supplies or platforms with differing speeds and capacities.

Many types of physical game pieces can be used to represent platforms, personnel, and supplies, as long as the pieces are easy to interact with, fit on the game map, and allow players to clearly distinguish between any relevant categories. It is possible to find playing pieces that look similar to what they represent (ships, trucks, people, boxes, barrels, etc.), which can be helpful when trying to identify pieces on a crowded game board. An example of a set of game pieces, along with what each piece represents, is shown in Figure 2.3.

FIGURE 2.3
Notional Set of Game Pieces and What They Represent

Transportation assets

1 game piece = 2 semi-trucks = 50 pallets

Personnel

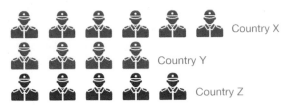

1 game piece = 2 truck drivers

Supplies

1 game piece = 25 pallets of Class V supplies

Miscellany

In keeping with the principle of designing flexible rules, it is useful to have a variety of other playing pieces that can be used to keep track of important information that could change as the game is played. When determining which additional pieces are required for a given game, it is helpful to ask the following: What information could be revealed over the course of the game that the players and adjudication team will need to incorporate into gameplay or game results? And how might the information depicted on the game maps change over the course of the game? The following are additional game pieces used in a previous iteration of this game:

- indicators tracking increases in time needed to unload supplies at spokes
- indicators showing that assets, supplies, or spoke locations were attacked
- pieces for reducing or increasing demand shown at spoke locations.

Players may also benefit from having informative resources for referral either during the game or when preparing for the game. These resources could include a primer on the game's scenario, a fact sheet on various entities in the game, or an overview of guidance from leadership. Resources should provide players with enough information to feel prepared prior to the game and to make decisions during the game. However, read-aheads should avoid overwhelming the players with detail, and any information provided should be covered on game day; there is no guarantee that busy players will read the materials in advance, much less study them in depth.

CHAPTER 3

Game Customization and Preparatory Analyses

As discussed in Chapter 1, a network logistics game of the type described in this report is suitable for exploring a variety of logistics problems. This chapter describes the process for designing game rules and materials. There are five primary components of the design process: developing the game scenario, determining how to represent demand across a network of locations, calculating the number of transportation resources that will be available to players, creating customized game rules, and conducting supplementary preparatory analyses. These components should be thought of as interlocking processes rather than sequential stages of design, and completing them will likely require some iteration. Note that, while the game scenario is a component of the game, it is discussed in this chapter rather than Chapter 2 because Chapter 2 focuses primarily on the physical components of the game. Therefore, we have integrated our discussion of the characteristics of a game scenario into our discussion of scenario development.

Develop Game Scenario

The scenario development step in game design is an essential part of developing most wargames, including this one. Care should be taken to create scenarios that are relevant and provide enough context to help players frame their decision space. One of the first obstacles in game play is players "fighting the scenario," which can be limited by providing necessary context via reasonable scenarios.[1] Scenario development provides the scope and detail necessary to develop the other game components, so it should be considered from the beginning of game design. For example, the logistics network cannot be modeled until the scenario specifies where the game is set geographically, what time frame the game's turns are intended to cover, what is driving demand for supplies, and which transportation resources are relevant to the game, among other things.

In this game, the scenario both directly and indirectly provides the contextual information required for the players to make intelligent decisions about protecting (or interfering with) transportation assets and supplies, as well as about which demands to meet first at which locations. The scenario directly shapes Blue and Red decisions by informing them about the nature of the actors they are portraying, allowing them to better consider what actions they might take. It indirectly shapes decisions about resource allocation because the expected level of activity at a location (and thus its approximate importance) is reflected in the demand at that location.

In general, the scenario should describe the crisis or conflict creating demand for supplies, and updates should be provided periodically during the game. However, the scenario should not go into detail regarding the disposition of forces and resources unless the information is directly relevant to logistics (e.g., relates to attacks on transportation assets, destruction of supply storage facilities, or damage to transportation infrastructure). This helps mitigate the risk that players will spend time debating the merits of the overall operational scenario rather than the logistics-related decisions they are there to address. Even if detailed scenario information is not originally provided, players might request information about which geographic regions should be prioritized. Adjudicators should refrain from providing this information. Instead, they can explain the general CONOPSs being implemented in the broader scenario (if it is a wartime scenario) and note that activity at each location roughly corresponds to the demand shown on the game maps. Adjudicators should then have a conversation with the players about the different options they see for allocating resources and the trade-offs they are considering when weighing those options. Understand-

---

[1]   For more information on scenario design, see Perla (1990, pp. 203–211) and DeWeerd (1973).

ing how players make these prioritization decisions is often a key part of answering the game's research questions; this information is lost if the adjudication team makes these decisions for them.

For this game, the following elements will be particularly important for informing the decisions players will make during the game:

- *Information about the general progression of the crisis or conflict that is creating demand for supplies across the logistics network.* As noted above, this information should be provided at a relatively high level unless it relates directly to the logistics system. It can be based on such data sources as the outcomes of past operational wargames or historical events, or it could be developed to intentionally strain certain parts of the logistics system. In general, it will require less work to use scenarios that are familiar to the players because this both increases the data sources available to the design team and reduces the amount of information that needs to be provided to the players (because they bring their own expertise to the game).

- *Information about the nature of the adversary or environmental factors creating obstacles to supply movement.* This information will influence how players choose to prevent or mitigate threats and should be in line with the information players would have in a real crisis (e.g., Blue team players may have a general idea of an adversary's capabilities and overall goals but would not know how the adversary would employ their capabilities in a specific situation). The Red team should typically receive more detailed information on their own capabilities and goals than the Blue team.

- *Information about transportation platforms and operators available to move supplies,* including where the platforms and operators will be drawn from, any characteristics that might affect how players would choose to employ them, and whether they will be constrained to specific starting locations at the beginning of the game. This information will inform player decisions about how to safeguard supply movements and allocate assets across the logistics network.

Lay Out Network and Demand Representation

The purpose of this component of game design is to generate a detailed network diagram displaying routes between hubs and spokes, transit time and distance, and demand at spokes, as well as a geographic map displaying a summary of the information on the network diagram. Figure 2.2 from Chapter 2 is one example of a network diagram, meant for use by the adjudication team. Figure 3.1 identifies the key elements on this diagram. Figure 2.1 shows a corresponding geographic map, meant for player use. See Table 3.1 at the end of this section for a summary of the primary parameters that can be adjusted when making the network diagram, other aspects of the game affected by those parameters, and key considerations when setting parameter values.

The process for transforming information about the logistics system into workable game maps is described below. This section also provides exam-

FIGURE 3.1
Key Elements of Detailed Network Diagram

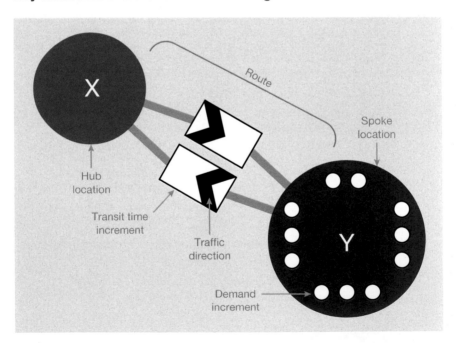

NOTE: Generated by RAND and not based on real data.

ples throughout of what the design process would have looked like for the hypothetical game described in Chapter 1.

1. *Identify hubs, spokes, and routes*: The first step in designing the network diagram is to determine which locations are relevant to the game, then determine which are hubs, which are spokes, and which are both. The scenario should provide the necessary scope and context to make these decisions. Use this information and any other applicable guidance on the scope of the game to determine which hub-spoke routes will typically be used by the transportation assets under consideration. For example, if the game is based on a certain CONOPS, use the locations and routes called out in that concept to create the base network for the game map. For instance, players could utilize the existing network of overseas bases operated by the U.S. Department of Defense as their starting point.[2] The goal is to identify the standard routes to depict on the game map, not to account for every possible route players could choose to take. Including numerous deviations from the standard routes on the game map can clutter it; these deviations can be handled in other ways if necessary during the game (discussed later in this chapter). Depending on the transportation network, it may also be useful at this point to identify the direction of travel along different routes.

Hypothetical game: As part of the scenario development process, the design team identified existing CONOPSs laying out the process of moving supplies during the potential conflict on which the scenario was based. These CONOPSs specified the locations of major supply distribution centers in the theater and supply storage facilities nearer to the frontlines that supplies would need to be transported to. Using the hub and spoke locations identified in the CONOPSs, information in the CONOPSs about how command of logistics would be divided geographically, and a map of terrain and major roads in the theater, the design team determined which hubs would send supplies to which spokes.

---

[2] For detailed lists of locations, as well as potential alternative locations to consider, Lostumbo et al. (2013) remains an excellent reference but should be updated by the game designer to reflect current posture and/or alternative CONOPSs.

2. *Measure route length*: Determine how far transportation assets will have to travel to move along each route. Use the actual path that the vehicles, vessels, or aircraft under consideration would have to travel for this measurement, rather than the straight-line distance between starting and ending points.

> *Hypothetical game:* The design team used geospatial software to determine the quickest route between each hub-and-spoke pair. This software also provided the length of those routes. During this process, the team also considered whether the roads along each route were appropriate for moving the trucks used in the game scenario, choosing alternate routes if necessary to avoid, for example, dirt or gravel roads.

3. *Determine average demand anticipated at each location*: The next step in building out the game is to determine how much demand players need to fill per turn at each location. Demand stems from the consumption rates of the units (or organizations) at each location for the type of supply considered in the game. Unit consumption rates over time, combined with information on the number and characteristics of units at each location, can be used to calculate demand over time at a location. While the appropriate source for demand data will depend on the scenario chosen, these data will typically come from operators in the form of information about the expected materiel consumption over time for units included in real plans or results from previous operational analyses. Standard planning factors can also be used, which can often be found in official, unclassified guidance on how to conduct analyses.
   – The game is simplest to design and execute when considering only a single type of supply (e.g., diesel fuel). If multiple types of supply must be included, it will likely be easiest to think of the logistics network as multiple networks overlaid on top of one another and complete the calculations described in this chapter once per each type of supply. Generally, the demand at each location should be broken out by type of supply on the network diagram and game map. That said, it may also be possible to simplify certain aspects

of the game. For example, if the same platform will transport all types of supply in the game, then the number of transportation assets available to the players needs to be calculated only once. If the same ratio of different types of supply is needed at every location, then the demand calculation could be completed in a single step and combined demand could be shown on the game maps, with the understanding that every "unit" of demand consists of, for example, 2 tons of Supply 1 and 1,000 gallons of Supply 2. In general, mixing classes of supply in a single game can be challenging unless significant simplification is possible because the game can quickly grow extremely complex at the expense of playability.

Hypothetical game: The logistics organization sponsoring the game had previously completed analyses determining the number of units operating from each spoke location and those units' anticipated supply utilization over time during the conflict covered by the scenario. The design team aggregated unit supply utilization to get expected daily supply utilization for the spoke location. This utilization rate was relatively stable, and analyzing players' response to changes in demand was not a primary focus of the game, so the research team chose to average daily utilization at each location over the first week of the conflict. After choosing an initial turn length of two days, the team multiplied the average daily utilization rate by two to obtain the demand per turn at each location.

4. *Identify characteristics of transportation assets:* Determine the speed and capacity of relevant vehicles, vessels, and/or aircraft. The simplest type of game will use a representative transportation asset with a single capacity and speed, but it is possible to run the game with multiple classes of transportation assets. Introducing multiple transportation assets will increase the complexity of both gameplay and adjudication, so it is best to differentiate between transportation assets only if the differences between them are substantial and maintaining granularity is necessary to meet game objectives.

Hypothetical game: The game sponsor informed the project team that they would likely use a specific type of military truck in the game scenario; they were able to provide specifications for vehicle dimensions and speed. They also informed the design team that the supplies being moved would typically be packed on standard-sized pallets. Using the truck and pallet dimensions, the design team calculated the maximum number of pallets that could fit on a single truck.

5. *Calculate transit time along each route:* Using asset speed and route lengths, calculate transit time for each route. This conversion is made because time is typically more operationally relevant than distance and because doing so makes it easier to determine how many turns it will take to complete a route. However, it may be more appropriate to continue representing routes in terms of distance rather than transit time if transit time will vary between different types of transportation assets. If distance is used, the following steps may require some adjustment.

Hypothetical game: By comparing the maximum speed of the trucks and the maximum speed limits along the routes chosen, the design team determined that the limiting factor on vehicle speed would be the characteristics of the roads, not the trucks, so the route transit times calculated earlier using geospatial software were used.

6. *Represent demand and transit time using discrete increments:* On the network diagram, demand per turn for each location and transit time for each route are displayed in discrete increments. The number of demand increments shown for each location is found by setting an initial value for the amount of demand represented by one increment and dividing demand at each location by that increment. The number of transit time increments shown for each route is determined similarly, with increments representing a specific length of time. The initial increment sizes chosen may require adjustment as

part of the iterative design process.[3] However, the size of the demand increments should be constant across locations. Likewise, the length of the transit time increments should be constant across routes. It is typically most convenient to round to the nearest whole number of demand or time increments when representing the results of these calculations on the game map, unless doing so would substantially misrepresent demand or transit time. For simplicity, displaying a fraction of an increment on the map should be the exception, not the rule.[4]

> *Hypothetical game:* The design team initially set the demand increment equal to the smallest two-day demand at any of the spoke locations—four pallets of supplies. Using the demand at each location and volume of supplies on a pallet, the team calculated the amount of demand increments required at each location, rounding to the nearest whole number. The team set the transit time increment size equal to half of the shortest transit time between any hub-and-spoke pair, then used this to calculate the number of transit time increments that would represent transit time along each route, also rounding this to the nearest whole number.

7. *Calculate distance that transportation assets can move each turn:* Use turn length and transit time increment size to determine how many increments (i.e., spaces on the network diagram) an asset can move per turn. Note that the speed of a transportation asset is already accounted for when using transit time. If using route distance, this calculation will need to account for speed. Regardless, rounding to the nearest whole number is generally necessary. If

---

[3] Unless otherwise specified, *increment size* refers to the length of time or amount of demand represented by an increment, not to the physical dimensions of the space used to represent that increment on the game board.

[4] A previous iteration of this game conducted by RAND researchers used transportation assets with a single speed, so it made sense to let each increment represent asset movement over a given time period. For games using transportation assets with notably different speeds, it may be easiest to instead think of each increment as a distance interval and allow different assets to move a different number of increments each turn.

doing so will eliminate needed granularity, decrease the size of the time increments.

Hypothetical game: The design team divided the turn length by the length of the transit time increments to determine how many increments a truck could move each turn.

---

8. *Iteratively adjust turn length, demand increment size, and transit time increment size while creating the transportation network diagram:* Use the previous calculations to create a network diagram displaying the calculated number of demand increments at each location and transit time increments along each route (see Figure 2.2 for an example). Adjust the length of each turn and the size of the demand and transit time increments until the number of demand increments displayed on the diagram for each location, the number of transit time increments shown along each route, and the number of transit increments each vehicle can move per turn are reasonable. This requires some judgment and may require iteration as the game map is designed. Table 3.1 summarizes how different design parameters interact, notes key considerations when determining their values, and provides some suggestions for weighing trade-offs during the design process.

Hypothetical game: A network map was created using the initial calculations for the number of demand increments at each location and transit time increments along each route. After this was done, it was clear that the longest routes had too many transit time increments to depict on a physical map that would fit on a conference room table. Because of this, the transit time increment size was doubled and the number of transit time increments for each route was recalculated.

9. *Depict demand on a geographic map*: Once the hub-and-spoke network and the demand at each spoke are finalized, create a geographic map of the region of interest with any notable features that could be relevant to players. Then, show the various hub and spoke locations on this map, and list the demand at each spoke location (see Figure 2.1 for example). It may make sense to represent demand via notecards or tokens placed on the map at the appropriate locations rather than printing this information directly onto the map. Doing so allows demand to be updated after the map has been printed to reflect players' decisions during gameplay or last-minute adjustments to calculations during game design.

Determine Supply of Transportation Assets and Operators

At the beginning of the network logistics game, players are given some number of transportation assets and/or operators to allocate along different routes. From a game design standpoint, this process is less complex than determining how to represent demand and transit time across the logistics network. It will be heavily dependent on the type of data available, so this section simply outlines some basic steps.

First, the number of transportation platforms and operators available in the game scenario should be determined. For operators, this will typically involve determining what characteristics operators must have to operate the transportation platforms included in the games. These characteristics could include certification status, nationality, clearance level, and membership in a particular organization (e.g., military unit), among other things. Much of this information can likely be provided by whichever organization is in charge of logistics in the game scenario. Once the relevant characteristics are determined, the total population of operators meeting those characteristics should be calculated. Example sources for this information include industry data on the number of commercial operators certified to work on specific platforms and government data on the number of mili-

TABLE 3.1

Summary of Network Diagram Parameters

Parameter	Parameter Impacts	Key Considerations	Recommendations
Transit time increment size	• Number of increments moved by a platform in a turn • Ability to differentiate between routes • Visual appeal of network diagram game board • Network diagram game board size	• For a given route, increment number increases as increment size decreases; too many increments can clutter the game board and increase its dimensions (increment dimensions on the board should not be smaller than the dimensions of the game pieces)	• Avoid making this value larger than the length of the shortest route • Make this value small enough that rounding off the number of transit time increments for each route does not eliminate needed granularity • Make this value large enough that, when the physical size of each increment on the network diagram is made larger than the size of the game pieces, the physical network diagram is not so large that players and adjudicators cannot easily view or reach parts of the diagram and it does not fit on available tables in the game venue
Demand increment size	• Ability to differentiate between spokes • Visual appeal of network diagram game board	• Platform capacity should be some multiple of this value • For a given location, increment number increases as increment size decreases; too many increments can clutter the game board • If rounding the number of demand increments required at each spoke results in an unacceptable loss of granularity, this value can be decreased	• Avoid making this value larger than the smallest demand at a spoke (it is possible to represent demand for partial demand increments on the network diagram, but this should be the exception, not the rule) • Use as large an increment size as is possible without losing granularity needed to answer the game's research question; this will make the network diagram easier to understand and will speed adjudication

Table 3.1—continued

Parameter	Parameter Impacts	Key Considerations	Recommendations
Turn length	• Number of increments moved by a platform in a turn • Time frame covered by game • Per-turn demand at spokes • Time required to play a game covering a specific time frame	• Practical concerns often limit the time available to play a game, and each turn can take a couple hours to play through, so this value typically determines the time frame covered by a game • Making this value too large can remove player decision points (e.g., players cannot respond to in-transit obstacles if platforms complete most routes in a single turn) • Making this value too small can prevent the game from progressing far enough to gain necessary insights in the time available to play it (e.g., effect of player decisions may not have time to proliferate across the network) • As this value increases, so do the demand at spokes and the number of increments moved by platforms, increasing adjudication complexity and time requirements	• Use a smaller turn length and play more turns to provide multiple opportunities for Red and Blue to interact with the logistics system; however, ensure that the turn lengths are long enough for the game to cover the required time frame given the maximum amount of time available to play it and the minimum time in which a turn can be completed • Avoid making this value so large that many transportation platforms need to reach most locations to meet demand in a turn; this makes adjudication less time-consuming • Increasing this value increases the number of transit time increments that platforms traverse in a turn; make it long enough that players feel like they are making progress, but ensure platforms cannot completely traverse every route in a single turn to provide Red opportunities to interfere with Blue platforms in transit and Blue opportunities to respond (e.g., by rerouting assets)

NOTE: Additional information used to create the game board, such as hub and spoke locations, route lengths, platform capacity, and spoke demand rate (independent of turn length), should typically be based on the scenario and background research.

tary personnel assigned to relevant missions.[5] Finally, the total number of potential operators should be cut down to reflect additional responsibilities that those operators may have in the scenario. For example, pulling all commercial operators from their typical jobs could have significant economic consequences. Likewise, it is probably unrealistic to assume that all of the military operators in one theater would be pulled out of that theater to support operations in another.

Determining the number of available transportation platforms available should follow a similar process. Required platform characteristics could include things like capacity, ownership, and hardening. As with operator eligibility requirements, this can likely be determined through discussions with the organization in charge of logistics in the scenario. The total number of military platforms can likely be provided by government data sources, whereas determining the number of eligible commercial platforms could require obtaining data from a market research firm. As with operators, the actual number of platforms available for the game will likely be smaller than the global number of eligible platforms. Like operators, any commercial platforms could be tied up in work critical to the health of the economy. Platforms may also typically operate outside the region where the game is set. Depending on their movement speed and the amount of warning players would have reasonably had about the events of the game, many of these platforms might be unable to reach the game's location in time to be relevant.

Second, transportation assets and operators are represented by physical game pieces when playing the game, so a key part of game design is determining how many assets or operators a given piece represents (this should be an integer). Determining the number of transportation assets represented by a single game piece requires having some idea of the level of demand at each location; if a game piece represents multiple transportation assets, the amount of demand that can be fulfilled when a game piece arrives at a location should generally not be much greater than the total amount of demand at any given location (because, for example, it is unrealistic to assume that

---

[5] Governments and trade associations frequently offer statistics about critical industries, which can be a potential source of data. For example, the Federal Aviation Administration tracks data on airmen (Federal Aviation Administration, 2023).

operators would send three trucks of supplies to a location where only one was needed, rather than sending the unneeded trucks elsewhere). If this is an issue, it can be rectified by letting one game piece represent a single asset or, if that fix has already been implemented, by increasing turn length, and thus demand per turn (see previous section).

Determining the number of operators represented by a game piece is simpler: It will generally be tied directly to the number of operators required for a given transportation asset. In the simplest case, the number of operators represented by a game piece will be the number of operators required for the transportation assets represented by a single game piece. If one piece represents multiple transportation assets or there are meaningful differences between operators, it is also possible for each operator piece to represent some fraction of the total requirement so that multiple operator pieces are matched with one transportation asset game piece. Whether there are meaningful differences between operators will depend on the choices that players are asked to make in the scenario. For example, if players are asked to consider a scenario in which Red could infiltrate of the pool of operators, Blue players may apply different vetting procedures to different groups of operators. Being able to specify that a platform is operated by people with varying levels of trustworthiness would be helpful in adjudicating how well Red can interfere with that platform via compromised operators.

Third, game design requires determining whether there are any meaningful ways to categorize transportation assets and operators, then making sure that the game pieces chosen can adequately distinguish between them. Relevant categories could include transportation asset type, capacity, and speed and operator training or nationality, among other things. The information required to make these categorizations can likely be obtained from the same sources used to determine how many operators and platforms meet eligibility requirements (e.g., industry reports or market research data). In a previous iteration of this game, vessels were categorized by flag and crews were categorized by nationality. This categorization was represented in the game via color-coding.

Finally, the number of transportation assets or operators represented by a game piece and the categorization scheme identified can be used to determine how many game pieces players should receive for each category. The data and analysis used for this can vary, but, in general, the number of game pieces to be used depends on how many of each category of asset or operator would be available for use given the scenario. Availability might depend on where the scenario is taking place; laws governing the credentials, certifications, or security characteristics of assets or operators; or other competing resource requirements (e.g., a subset of relevant transportation assets may be required to support a vital industry). It may also make sense to predetermine where pieces are in the transportation network when the game starts if that is not something players would have control over given the game's scenario.

Create Customized Game Rules

Game rules govern how player decisions are implemented and adjudicated by the adjudication team. In this game, Blue players are attempting to optimize the use of their transportation resources under conditions of uncertainty, and their decisions generally fall into two categories: decisions about allocating transportation resources to meet demand and decisions about preventing and mitigating unexpected obstacles. More specifically, Blue players are likely to do the following:

- Assign operators to platforms.
- Assign platforms to routes.
- Reroute platforms in transit.
- Develop strategies to protect operators, platforms, and/or supplies.
- Develop strategies to mitigate unexpected setbacks.

Red players will primarily focus on developing strategies to interfere with Blue logistics.

In this game, the movement of transportation assets and supplies is adjudicated according to formal rules, while adjudication of how players impose, prevent, and mitigate obstacles relies on SMEs. Therefore, the rules primarily specify how transportation platforms, operators, and supplies are allowed to move around the network diagram during the adjudication process. Because this game design can be adapted to a variety of contexts, the details of the rules used may differ somewhat from game to game. Table 3.2 lists the types of rules that will be required for most games using this design,

TABLE 3.2

Examples of Typical Game Rules

Game Rule	Example
Assigning transportation assets to routes	Transportation platforms and operators can be assigned to any route at the beginning of the first move, with their starting point at that route's hub location.
Assigning operators to transportation assets	Transportation operators can be assigned to any hub at the beginning of the game and may then be assigned to any platform leaving from that hub. The operators on a given platform can be switched only while that platform is at a transportation hub, and they must be switched with other operators at that hub.
Moving transportation assets	Each transportation platform and its associated operators and supplies can move three spaces on the network diagram in a turn. Each hub and spoke location count as once space.
Loading and unloading supplies at hubs and spokes	Any number of transportation platforms may reach the same hub or spoke location in a turn, and there is no limitation on the number of platforms that can be loaded at one hub location in a single turn; however, a maximum of four transportation platforms can unload supplies at the same spoke location during a turn.
Resetting demand met each turn	Unless noted on the game board, the quantity of supplies that reached a spoke location during a game turn is reset to zero at the beginning of the next turn.

along with an example of each; Table 3.3 lists additional rules that may be appropriate depending on the game's scenario and objectives. A narrative description of how both sets of rules were implemented in a previous iteration of this game can be found in Chapter 4.

The design of game rules should rely on research into how the various game components function. For example, the rules for choosing or adjusting routes for transportation assets should be faithful to the primary characteristics of how the process actually works. In the case of the game RAND conducted, SMEs were relied on. In other cases, written standard operating procedures or other data sources could be used. If transportation assets are typically assigned to new routes only once they return to their hub location, then transportation assets in the game should be rerouted only under the same circumstances. Likewise, throughput time and capacity at spokes should be reflected in how pieces are moved to and from those loca-

TABLE 3.3
Examples of Other Potential Game Rules

Game Rule	Example
Adjusting demand at spoke locations	If storage capacity is reduced at a spoke location or the game scenario indicates that demand has changed at that location, an appropriate number of demand increments on the network diagram will be covered up in consultation with SMEs.
Adjusting the number of available transportation assets or operators	If transportation platforms or operators are rendered unable to conduct transportation activities, the appropriate game pieces will be removed from the network diagram in consultation with SMEs.
Rerouting transportation assets	A transportation platform, along with its associated operators and cargo, may be rerouted at any time while in transit or otherwise sent along a route not shown on the network diagram, but it must spend an appropriate amount of time in transit to reach its new destination (refer to travel time table generated prior to game).
Displaying the effects of disruptions or obstacles	If unloading facilities at a spoke location are damaged such that the amount of time required to unload a transportation platform is increased, a numbered token indicating how many platforms can now unload at that location in a single turn will be placed at that location on the network diagram.

tions during the game. Note that game rules will never represent real processes with perfect fidelity—and that's not the point. The game rules are an abstract model of real processes, meant to reflect the most important dynamics and constraints of those processes and produce game results that are close enough to reality to generate useful insights.

Conduct Additional Required Analyses

While the primary game design processes are covered in the previous three sections, there are additional analyses that could be required depending on the scenario and objectives of the game. These include analyses required to adapt adjudication to creative player choices, analysis of the impact of obstacles or disruptions on the system, and analysis of the impact of strategies that players may use to prevent or overcome obstacles or disruptions.

Because implementing flexible rules is a key design principle of this game, analysis should be conducted to support adaptive adjudication where appropriate given the scope of the game. For example, if it is realistic for players to deviate from the standard transportation network, it is helpful to calculate the number of transit time increments between nonstandard hub-and-spoke combinations ahead of time. If players could decide to stockpile resources at certain locations before focusing on meeting demand at other locations, it is important to know how much storage capacity each location has.

Understanding how disruptions or obstacles can affect the results of the Blue team's decisions is an important part of ensuring that the game faithfully represents the scenario and meets its objectives. The types of analysis needed will be highly dependent on the types of disruptions or obstacles that will be encountered, and requires thinking carefully about which will be most likely given the scenario. If there will be an active Red team, the following are examples of potentially helpful analyses:

- impact of kinetic attacks on transportation assets, operators, and hub/spoke locations
- impact of cyber attacks on command and control capabilities
- impact of coercion on operators.

Determining what capacity an adversary will have to carry out various actions given the conditions laid out in the scenario is also important in order to present the Blue team with realistic challenges. While the Red team should be aware of their capacity limitations, it typically makes sense to withhold this information from the Blue team unless certain limitations would realistically be known to both sides.

Adversary actions are not the only possible obstacles to logistics. Other potential obstacles include adverse weather, natural disaster–related infrastructure damage, traffic jams, and worker shortages due to illnesses. If obstacles will arise from environmental and other passive sources, relevant analyses could include

- impact of traffic disruptions on transit time
- impact of illness on operator availability

- impact of bad weather on transit and loading and unloading times at hubs or spokes.

Understanding how players could prevent or mitigate the impact of disruptions may also be necessary. As with analyzing the impacts of disruptions, this may require predicting which courses of action players will find most feasible. In general, the impacts of both obstacles and prevention or mitigation strategies are not represented by formal game rules in this game. Instead, adjudication relies on expert judgements made both before and during the game. Because of this, including SMEs on the adjudication team during the game itself can allow adjudicators to realistically adapt to unexpected choices. These SMEs should ideally have deep expertise in the relevant domains necessary to answer the research question. For instance, if the game is about modeling the logistical requirements to move munitions, some set of experts in munitions, logistics, and air combat operations is necessary. Providing players with a set of available options could also reduce surprises, at the cost of limiting the game's ability to fully explore player insights.

CHAPTER 4

Gameplay and Adjudication

This chapter describes the process of how the game described in this report is played, with examples from a hypothetical game included throughout.[1] At the beginning of the game, the players are given a scenario that requires the Blue team to transport supplies to various locations shown on a map in order to meet demand. Members of the Blue team are told what type of resources they can use as they attempt to meet their objectives. If a Red team is playing, members of this team are also briefed on their objectives, which typically involve preventing or altering Blue supply movements. At this point, the game turns begin and proceed according to the following process:

1. The Blue and Red teams separate to plan their first move before communicating their intent to the adjudication team.
2. The adjudication team uses this intent to determine how assets and supplies move throughout the transportation network and resolve conflicts between Blue and Red actions. The adjudicators then calculate how well demand has been met at various locations and record relevant data.
3. The adjudication team briefs the players on Blue and Red team actions during the turn, amount of demand met at each location, and any other notable changes. If any information would not be apparent to either Red or Blue, the relevant team may be briefed separately.
4. The game board is reset as needed and players start planning their next turn.

---

[1] The results (i.e., outcomes) of the actual game run for this project are not available to the general public. This report incorporates our lessons learned about how to streamline game design and play throughout.

Steps 1–4 are repeated either a specific number of times or as many times as time allows. After an appropriate number of turns are completed, the adjudication team gathers the players together for a hotwash session during which participants (including adjudicators) discuss the results of the game and the insights they took from it. Each part of this process is described in more detail below. Note that this process reflects a previous iteration of this game that was conducted by RAND and may vary slightly for other network logistics games. One of the major sources of variation will be the nature of the Red team, which can consist of players or adjudicators and portray active or passive sources of obstacles to Blue logistics. For example, the Red team will not participate in the planning phase of each turn if Red actions are predetermined and executed by adjudicators instead of players; this could be the case in games designed to explore how the Blue team responds to environmental obstacles (e.g., natural hazards or traffic).

Orientation: Introducing Players to the Game

The first stage of gameplay is to introduce the players to the game scenario, describe the types of resources available to the players, and outline the rules governing how players can use those resources to solve problems posed by the scenario. The way in which players are introduced to the scenario and rules of the game can have an impact on player engagement. Ensuring that players buy into the game scenario and design before they start playing can help the rest of the game run smoothly. It may also be helpful to provide player aids detailing key aspects of the scenario. These aids could include maps (of military operations, major flooding, evacuation routes, etc.) or capability summaries (orders of battle, vehicle fact sheets, etc.).

Introducing the game scenario involves describing the logistics network and distribution of demand, as well as providing players with contextual information to use when balancing their priorities as the game progresses. For example, a contested operations scenario might provide details on the road to war, the nature of an ongoing conflict, and the type of operations being conducted from each spoke location. Providing this context encourages players to think carefully about how meeting or not meeting demand at various locations will affect the broader conflict rather than treating the game as an abstract optimization exercise. Important updates to this scenario are provided periodically, but they will focus primarily on how the broader scenario affects the logistics system. In this game, it is assumed for simplicity that the players' actions do not, broadly speaking, change the overall operational scenario. Once the players understand the scenario and associated demand for supplies across the network, they need to know what resources they will have available and whether there are relevant distinctions between different resource categories.

The final step for orienting players is to describe the game rules, emphasizing the decisions that players will be asked to make during the game. Because the game design allows players to operate at a lower level of complexity than the adjudication team, it is not necessary to explain in detail how the adjudication team will track resources and adjudicate player decisions, although a brief overview can be helpful. Instead, the briefing should explain to the Blue players that they need to determine their priorities for meeting demand across the network—that is, decide where meeting demand is critical and where it is not. The Blue players also need to describe how they would allocate resources and overcome obstacles to meet their objectives. Red players should similarly be instructed to develop a plan to further their goals in the game and describe how they would like to carry out that plan. Once the players have been introduced to the game (and throughout the introduction process), the adjudication team should confirm that the players understand what they are being asked to do and address any questions or concerns; with this complete, the game turns can commence.

Each Turn: Obtaining Player Decisions

Once the game has been introduced to the players, the first game turn begins. In the first part of each turn, the players discuss the problems that the game's goals, context, and progression have presented them with, brainstorm solutions, and debate the merits of those solutions before communicating final guidance regarding priorities and intent to the adjudication team. The geographic map is intended to aid players in this planning process by laying out the state of demand across the logistics network. Although the network diagram is intended primarily for use by the adjudicators, players can also consult it for a more detailed layout of the position of specific transportation assets. Depending on the game, there may be both a Blue team and a Red team tasked with making different decisions, as described below.

Blue Team

The goal of the Blue team is to successfully allocate resources such that demand is met across the logistics network; if meeting demand at every location is not possible, their goal is to successfully allocate resources to meet demand according to their strategy for prioritizing locations. While making these decisions, they also need to discuss how they would ensure the logistics system runs smoothly, considering, for example, how they would conduct command and control, ensure the reliability of assets and operators, or prevent Red team interference. The adjudication team should be listening to these conversations to understand what the Blue team would like to do and how they would like to do it. Although it can take multiple turns for platforms to traverse transportation routes, the Red team will generally provide new obstacles for the Blue team to respond to each turn. This both helps maintain player engagement throughout the game and provides a more rigorous stress test of the Blue team's ability to keep the logistics system functioning effectively.

Hypothetical Blue Turn: The Blue teams notes that most of the locations with high demand for supplies are in Country X. They also know, from previous games exploring broader military operations within the same scenario, that the most critical military activity will likely occur near the borders that Countries X and Y share with the adversary country. Based on this, they decide to prioritize fully meeting demand throughout Country X and at the spokes in Country Y that are closest to its border with the adversary country. The Blue team knows that the adversary country has a history of using special forces to sabotage transportation platforms and supply storage facilities, so Blue tells the adjudication team that they will increase the number of personnel guarding supply storage facilities and truck parking lots.

Red Team

In general, the Red team makes decisions about how to interfere with the Blue logistics system. A Red team representing an intelligent adversary will typically balance their play between three objectives: disrupting Blue by any means possible, playing according to their best understanding of actual adversary behavior, and playing in the way that best supports game objectives. The goals of the game drive which objective(s) are most important; this can, in turn, drive decisions about the structure of the Red team (Caffrey, 2019). There are two options for including a Red team in this game: using an independent Red team consisting of players or making the Red team a wing of the adjudication team. In games concerned with Red decisionmaking when faced with Blue actions, the Red team's focus will likely be on realistically portraying potential Red behavior. In this case, the Red team will typically be independent from the adjudication team. In a game more focused on providing a stress test for Blue concepts, Red team goals may be to disrupt Blue by any means possible or to challenge Blue in specific ways that support the game's objectives (or some combination of the two). An independent Red team might be most useful in creating unexpected challenges for Blue to overcome, while an integrated Red/adjudication team might more effectively act in ways supporting specific game objectives. The latter option is particularly appropriate if the obstacles will be environmental or otherwise passive, as opposed to the result of interference by an intelligent adversary. Regardless of the Red team's structure or objectives, the adjudication team needs to understand what actions the Red team plans to take and what they intend to accomplish by those actions.

Note that including SMEs on the Red team is particularly important because the Red team's decisions are much less constrained than the Blue team's in this game design. Whereas the Blue team is focused solely on moving a specific set of resources through a network, the Red team can choose a wide variety of targets and interference methods. If desired, the Red team can be constrained to a specific set of actions with well-defined rules for modifying Blue's movements, but this is not necessary, particularly if the Red team is embedded in the adjudication team and includes SMEs. This was the case for a previous iteration of this game run by RAND, in which the Red team, which had deep expertise on Red capabilities and their effects, made decisions in line with their understanding of likely Red behavior and adjudicated the impact of those decisions using their understanding of prior research.

Hypothetical Red Turn: Drawing on their knowledge of Red doctrine and Red and Blue actions in previous games using the same scenario, the Red team believes that the adversary they are representing would be most interested in interfering with the resupply of Blue forces near the border between the adversary country and Country X. They know the adversary has sophisticated special forces teams, so they tell the adjudicators they would send those teams to destroy spoke storage facilities nearest to the border. They also know that there are a small number of highway tunnels through which much of the traffic transiting Country X must pass and that the air defenses of Country X are likely insufficient to handle a large cruise missile attack, so they decide to use cruise missiles to attack highway tunnels concurrently with a missile attack on bases near the border.

Each Turn: Adjudicating Player Decisions

Once the players have described their intent to the adjudication team, the adjudication team translates that intent to detailed movements of transportation assets, operators, and supplies throughout the logistics network (as represented by playing pieces on the physical network diagram) according to the game rules. The rest of this section describes rules that are based on a previous iteration of this game and would be suitable for a variety of network logistics games, but it is also possible to add or adjust rules. During the first turn, operators and transportation assets are assigned to hub locations according to stated player intent, and operators are assigned to transportation assets. On every turn, each transportation asset (with its associated operators) is then moved a set number of transit time increments (calculated during game design) along its route on the network diagram, with each hub and spoke counting as one transit time increment. When a transportation asset reaches a spoke location, supplies are unloaded at that spoke according to the transportation asset's capacity. The number of transportation assets that can unload supplies at a given spoke during each turn is limited according to prior analysis of the length of time it would take to unload supplies and the number of transportation assets that can unload at one time. When a transportation asset reaches a hub location, it is automatically reloaded and can be assigned to any other route connected to that hub as needed to meet player intent. Most routes carry traffic in both directions, but some may be marked as one way only. If players want to move resources along a nonstandard route, the progress of these resources is tracked off map.

As the adjudicators move Blue game pieces around the game board, the Red team players embedded in the adjudication team will note where the movement of resources would be disrupted by Red actions and adjust the result accordingly. For example, they may remove assets that have been incapacitated from the game board. The Red team will also adjust characteristics of the logistics network based on Red actions, which includes noting where Red interference has changed demand for supplies at a spoke or restricted the number of transportations assets that can arrive at a spoke during one turn.[2] If demand has changed at a spoke, the geographic map should also be updated at this point to reflect that. Figure 4.1 illustrates the process of tracking the movement of transportation resources and supplies on the network map for one route. The supplies, platforms, and operators shown would be represented by physical game pieces during an actual game. Each subfigure shows the following:

- Figure 4.1a: This diagram shows the position of transportation assets and the demand met at the end of Turn 2 of a hypothetical game. In this turn, all of the demand at Spoke F (three units of supplies) was met, as shown by the black playing pieces sitting on all of the circles indicating demand at Spoke F. In this hypothetical game, platforms can move three spaces per turn, so another supply truck carrying three units of supplies is expected to reach Route F in the next turn as well.

---

[2] In some wargames, where the outcome of conflict is the object of study, it is not considered best practice to give asymmetric advances, such as perfect situational awareness, to one side or the other. Because the primary value of this game is to understand how Blue considers mitigating risks to disruption in its logistics system, having Red function as an adjunct of the adjudication team allows Red to focus on challenging different aspects of the system. That being said, the results adjudication does not have to assume that Red combat systems or outcomes also have perfect situational awareness.

FIGURE 4.1

Tracking Resource Movements on the Network Diagram

(a) State of Route D–F at end of Turn 2

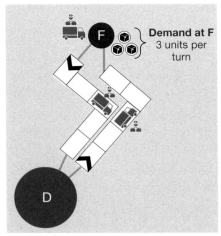

(b) Route D–F reset for beginning of Turn 3

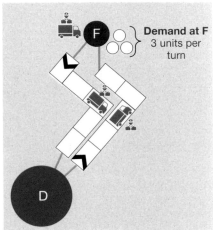

(c) State of Route D–F after resources are moved based on Blue plans for Turn 3

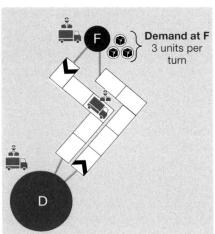

(d) State of Route D–F after resource movement is adjusted based on Red actions for Turn 3

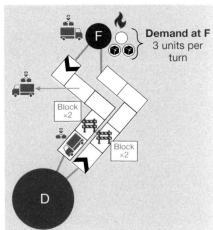

- Figure 4.1b: The game board has been reset for the beginning of Turn 3 by removing the playing pieces indicating that demand was met at Spoke F.
- Figure 4.1c: After the players have planned their actions for Turn 3 and conveyed their intent to the adjudication team, the adjudicators move transportation platforms and operators in line with Blue plans. If a platform reaches a spoke location, a number of black supply pieces equaling the platform's capacity are placed on the circles indicating demand. In this example, a platform reached Spoke F and unloaded three supply units.
- Figure 4.1d: This figure shows the location of resources and amount of demand met on Route D-F at the end of Turn 3. After transportation resources and supplies are moved according to Blue plans, their positions are modified based on Red actions. In this hypothetical game, Red chose to attack spoke supply facilities, successfully reducing the storage capacity at Spoke F by 33 percent and destroying an equivalent amount of the supplies that reached Spoke F. One supply piece was removed from the board and a token indicating that damage was done was added to the board to reflect this. They also used psychological operations to intimidate operators, persuading the driver of a truck leaving Spoke F to sabotage their own vehicle to avoid continued involvement in the conflict. Both the operator and platform pieces were permanently removed from the board and from the Blue team's pool of resources. Finally, Red targeted critical chokepoints along Blue's transportation routes with missile strikes, successfully blocking a tunnel along Route D–F, forcing trucks to take a detour that increases transit time along the route by 20 percent. The truck that had originally reached Hub D was moved back a space to reflect the delay and tokens indicating damage to the route and specifying the extent of the delay were added to the board.

Once all transportation assets have been moved and unloaded if necessary and all adjustments due to Red interference have been made, the number of demand increments filled at each spoke location is counted and reported to the players. Changes in the pool of available resources or the demand at spoke locations are also calculated and briefed to players, along with an account of the impacts of Red interference and any relevant updates to the broader scenario. If any results of Red or Blue activities would be unknown to the other team, those results are briefed to each team separately. After the results are briefed, players will ask whatever questions they require to clarify the state of the logistics system and begin to plan their next turn. The adjudication team will then reset demand met per turn on the network diagram back to zero. If demand at any location has changed, the geographic map is also updated. Note that the first game turn is often the longest and most arduous, as the players are getting used to the game concepts and rules; subsequent game turns should be significantly more straightforward.

Each Turn: Data Collection

Data collection occurs both while players make decisions and while those decisions are adjudicated. The type and amount of data collected will depend on the purpose of the game, but most games will at minimum record player decisions and the amount of demand met at each location. The latter may be sufficient "scorekeeping" for games played for education or entertainment purposes. This game design can be used for a variety of purposes, including exploring the feasibility of concepts, generating solutions to problems, discovering system vulnerabilities, and identifying patterns of decisionmaking. For these types of games, additional data collection will be necessary. Table 4.1 provides examples of data that could be collected to answer different research questions.

TABLE 4.1

Data Collection to Answer Research Questions

Potential Research Question	Useful Data to Collect
Does a given logistics CONOPS enable delivery of enough supplies to meet minimum demand across the network in a contested environment?	• Amount of demand met at each location over time • Demand at each location over time (if demand is flexible) • Supply of transportation assets and operators over time (if attrition is a possibility)
What strategies can be used to minimize the impact of logistics disruptions due to unexpected obstacles?	• Blue team discussions—especially their priorities, strategies, and intent for each move • Red team strategies and uncovered vulnerabilities in Blue concepts and/or systems (if playing a contested game) • Results from adjudicating the impact of Red interference • Amount of demand met at each location over time
What vulnerabilities can an adversary exploit that could significantly reduce Blue ability to meet demand across the network?	• Red team strategies and uncovered vulnerabilities in Blue concepts and/or systems (if playing a contested game) • Results from adjudicating the impact of Red interference • Amount of demand met at each location over time
How do human decisionmakers respond to unexpected setbacks and shortfalls in their ability to meet demand across the network?	• Blue team discussions—especially their priorities, strategies, and intent for each move • Results from adjudicating the impact of Red interference • Amount of demand met at each location over time • Supply of transportation assets and operators over time (if attrition is a possibility)

Hypothetical game: Because the game was designed to think through vulnerabilities in the Blue logistics system, the data collection team records the following:

- how Blue decides to protect their resources and the considerations they weigh in making that decision
- which parts of the logistics system Red attacks and why
- the results of adjudicating the interactions between Blue defensive measures and Red interference
- how Red and Blue respond to the actions of the other team in previous turns, including whether the other team's actions were surprising or particularly difficult to deal with
- how much demand is met at each location over the course of the game
- how Blue's supply of transportation resources is changed by Red actions over the course of the game.

Hotwash: Discussing Lessons Learned

Many games end with a conversation between the players and the adjudication team about the results and key takeaways from the game. This conversation is an opportunity for participants to voice what they learned about the problems they faced and for the group to synthesize individual comments into a more cohesive set of insights. For instance, it may be appropriate to ask what challenges the players faced during the course of the game that surprised them. Alternatively, it might be appropriate to ask what players might have done differently in a second iteration of the game. It is important to bear in mind that these are examples only, and that the specific questions asked should be aimed at eliciting the necessary feedback from the players to help meet the objectives of the game. This time can also be used to identify how well the game functioned—whether players felt constrained by the rules and had issues with how well it represented real problems or whether players had the space to explore reasonable strategies and learn useful things from their experience.

CHAPTER 5

Conclusion

The game described in this report can be used to study a variety of scenarios and logistics networks, lending itself particularly well to scenarios in which resources are limited or unexpected obstacles could appear. A previous iteration of this game considered movement of military supplies in a contested scenario, but other scenarios could include moving casualties through the military medical system during a large-scale conflict—a scenario with strong competition for resources and high time pressure—or moving relief supplies to affected communities after a natural disaster—an uncontested scenario with environmental and passive obstacles, such as dangerous weather, degraded communications, and damaged transit infrastructure. The game design described can be useful as long as the scenario involves using resources to move supplies through a relatively fixed network and there are uncertainties or problems in doing so that players can grapple with during the game.

Potential uses for this game include exploring new concepts of logistics operations, illuminating vulnerabilities in logistics networks, understanding the dynamics of supply movements through a logistics network, highlighting the broader impacts of logistics planning, and brainstorming solutions to challenges. Exploring new CONOPSs using a game like this is helpful because the game's artificial environment provides a relatively cost-effective and efficient way to gain a basic understanding of the implications of implementing future capabilities and radical strategies prior to making real-world investments. The simplified model of reality represented within this game also facilitates examination of the logistics system itself. As players attempt to meet goals and overcome obstacles, they will likely encounter challenges and unexpected outcomes. By playing the game, they can better understand how the effects of their decisions propagate throughout the complex logistics system, as well as which aspects of the system are particularly vulnerable to disruption.

Players' success or failure in meeting their goals during the game can also be a valuable starting point in understanding how logistics affects operations more broadly. Operational wargames focusing on combat often assume that supplies will be available when needed, and this game could be used to identify scenarios and conditions under which that assumption does or does not hold true and allow for the exploration of the operational impacts of shortfalls. If the game scenario presents players with a problem they find themselves unable to solve under the game's base conditions, it may also be appropriate to use the game as a means to discuss what could be improved about the logistics system. Because this design lays out different aspects of the logistics system and scenario in a tangible format, players can manipulate game pieces and experiment with using different strategies to identify where they are facing shortfalls. Perhaps they have too few vehicles, are trying to move too many supplies to too many locations, or cannot sufficiently mitigate obstacles imposed by the Red team. Players can use these insights to brainstorm solutions and inform future analysis.

When designing future games, there are a few potential modifications that might be appropriate. First, this game design could be modified to accommodate networks that are more interconnected than the hub-and-spoke network shown in this report. Second, game designers could create more detailed mechanisms governing how the Red team can interfere with supply movements and how such interference is adjudicated, which could be helpful in a game exploring the behavior of the Red team in response to Blue supply movements. Third, this game could be adapted so that both Red and Blue are attempting to conduct their own logistics while disrupting their adversary's logistics. Finally, this game could be integrated into a series of other wargames. For example, a political-military game in which the Red and Blue teams compete for influence in other countries could inform the set of transportation assets that players start the logistics game with, or the amount of demand players meet in the logistics game could constrain Blue team operations in a combat-focused operational game. Alternatively, expansions of this game could help policymakers in the U.S. Department of Defense understand the logistical supportability of alternative warfighting CONOPSs.

As described above, this game design is flexible enough to accommodate a variety of scenarios, objectives, and modifications. However, future iterations of this game should preserve the simple, transparent, and tangible way in which logistics networks are represented within the game. Game components and rules should represent the logistics system in a way that is abstract enough for curious players to easily grasp while providing enough detail to distinguish between relevant factors and represent reality with an appropriate level of fidelity. Players should also be enabled to engage with the problems presented by the game at whatever level of detail they are most comfortable with. Finally, the game design should be flexible enough to allow players to explore innovative solutions to the problems they have been given. Maintaining these key strengths of the game design will allow it to remain a useful and impactful tool, even as it is adapted to answer new questions and explore new possibilities.

References

Army Doctrine Publication 4-0, *Sustainment*, Department of the Army, July 31, 2019.

Bartels, Elizabeth M., *Building Better Games for National Security Policy Analysis: Towards a Social Scientific Approach*, dissertation, Pardee RAND Graduate School, RAND Corporation, RGSD-437, 2020. As of November 3, 2022:
https://www.rand.org/pubs/rgs_dissertations/RGSD437.html

Caffrey, Matthew B., *On Wargaming: How Wargames Have Shaped History and How They May Shape the Future*, Naval War College Press, 2019.

DeWeerd, Harvey A., *A Contextual Approach to Scenario Design*, RAND Corporation, P-5084, September 1973. As of January 25, 2023:
https://www.rand.org/pubs/papers/P5084.html

Donnelly, R. Hank, Marc E. Garlasco, Daniella N. Mak, Kara Mandell, Ed McGrady, Justin Peachey, AnneMarie Mandazzo-Matsel, Rebecca Reesman, and Christopher Ma, *LOGWAR 15: Analysis Report*, CNA, 2016.

Federal Aviation Administration, "U.S. Civil Airmen Statistics," webpage, last updated January 19, 2023. As of January 25, 2023:
https://www.faa.gov/data_research/aviation_data_statistics/
civil_airmen_statistics

Hanley, John Thomas, Jr., *On Wargaming: A Critique of Strategic Operational Gaming*, Ph.D. dissertation, Yale University, 1991.

Krievs, Daniel A., *Integrating Agile Combat Support Within Title 10 Wargames*, Air Force Institute of Technology, 2015.

LaPlante, John B., David P. Garner, and Patricia Insley Hutzler, "Logistics in Wargaming: An Initial Report," *Joint Forces Quarterly*, Winter 1995–1996.

Lostumbo, Michael J., Michael J. McNerney, Eric Peltz, Derek Eaton, David R. Frelinger, Victoria A. Greenfield, John Halliday, Patrick Mills, Bruce R. Nardulli, Stacie L. Pettyjohn, Jerry M. Sollinger, and Stephen M. Worman, *Overseas Basing of U.S. Military Forces: An Assessment of Relative Costs and Strategic Benefits*, RAND Corporation, RR-201-OSD, 2013. As of January 25, 2023:
https://www.rand.org/pubs/research_reports/RR201.html

Mays, Robin L., Nicholas L. Chapman, Lauren R. McBurnett, Colton D. O'Connor, Thomas Humplik, Eric V. Heubel, and Christopher K. Ma, *Advancing Globally Integrated Logistics Effort 2017 Wargame Report*, CNA, 2017.

Mueller, Karl, "Filling the Gap," *BATTLES Magazine*, No. 11, July 2019, pp. 53–57.

Perla, Peter P., *The Art of Wargaming: A Guide for Professionals and Hobbyists*, U.S. Naval Institute, 1990, pp. 203–211.

Perla, Peter P., and E. D. McGrady, "Why Wargaming Works," *Naval War College Review*, Vol. 64, No. 3, 2011.

Rubel, Robert C., "The Epistemology of War Gaming," *Naval War College Review*, Vol. 59, No. 2, 2006, pp. 108–128.

Vershinin, Alex, "Feeding the Bear: A Closer Look At the Russian Army Logistics and the Fait Accompli," *War on the Rocks*, November 23, 2021.

Weuve, Christopher A., Peter P. Perla, Michael C. Markowitz, Robert Rubel, Stephen Downes-Martin, Michael Martin, and Paul V. Vebber, *Wargame Pathologies*, Center for Naval Analyses, September 2004.